EBOOKS AND EDITORS:
WHAT YOU NEED TO KNOW

KEVIN CALLAHAN

THE-EFA.ORG

Copyright © 2020 by Kevin Callahan
Cover and design © 2020 Editorial Freelancers Association
New York, NY

All rights reserved.
No part of this publication may be reproduced, distributed, or transmitted in any form or by any means, including, but not limited to, photocopying, recording, or other electronic or mechanical methods, without the prior written permission of the publisher, except in the case of brief quotations embodied in critical reviews and certain other noncommercial uses permitted by copyright law. For permission requests, write to the publisher at "Attention: Publications Chairperson," at the address below.

266 West 37th St. 20th Floor
New York, NY 10018
office@the-efa.org

ISBN: paperback 978-1-880407-44-8
ISBN: ebook 978-1- 880407-45-5

Callahan, K. *Ebooks and Editors: What You Need to Know*

Published in the United States of America by the Editorial Freelancers Association.
Subject Categories: LANGUAGE ARTS & DISCIPLINES / Publishers & Publishing Industry | DESIGN / Book

Legal Disclaimer

While the publisher and author have made every attempt to verify that the information provided in this book is correct and up to date, the publisher and author assume no responsibility for any error, inaccuracy, or omission.

The advice, examples, and strategies contained herein are not suitable for every situation. Neither the publisher nor author shall be liable for damages arising therefrom. This book is not intended for use as a source of legal or financial advice. Running a business involves complex legal and financial issues. You should always retain competent legal and financial professionals to provide guidance.

EFA Publications Director: Robin Martin
Copyeditor: Cheryl Murphy
Proofreader: Robin Martin
Book Designer: Kevin Callahan | BNGO Books
Cover Designer: Ann Marie Manca

Contents

One for all, and all for one!

Introduction

Acknowledgments

What's in this booklet?

Chapter 1: What is an ebook?

Chapter 2: What is EPUB?

Chapter 3: What's the workflow?

Chapter 4: Semantic content matters

Chapter 5: Expand content

Chapter 6: Hyperlinks

Chapter 7: Which italic is this?

Chapter 8: Tables, math, and science

Chapter 9: Ebook design

Chapter 10: Color and images

Chapter 11: Ebook QA

Appendix A: XHTML and CSS samples

Appendix B: Accessibility checklist

Appendix C: Transmittal to the ebook developer

Appendix D: Resources

Glossary

About the Author

About the Editorial Freelancers Association (EFA)

If you're going to be working with ebooks*, read ebooks. Lots of them.

Learn how devices work.

What features do you like?

What drives you nuts?

Notice how you read.

**Whether you know it or not, most books you edit
will be published as an ebook as well as print.
So, everyone should take the plunge
and read an ebook or two.*

One for all, and all for one!

It's not just for the Three Musketeers anymore.

Everyone who works on a manuscript adds a layer of expertise. The development editor shapes the book; the copy editor refines the language; the designer fashions the presentation; the proofreader corrects mistakes. It's a group effort. While copyediting in the solitude of your home office, you're marking head-level hierarchy, bold, italics, and numbered lists. You're querying the author on language choice; revealing the structure for the designer, typesetter, and ebook developer; and building a stylesheet for the proofreader to check against.

We're on the same team making books for multitudes of readers: travelers picking up a paperback at the airport, library patrons borrowing ebooks, and blind readers using text-to-speech functions in e-reader software. The more we understand each other's roles, the better our work will be.

Introduction

I recently saw a Twitter conversation about how annoying reading an ebook can be: how poorly some devices display content, how inconsistent they are in following book-making standards, how uneven the experience is from device to device. One tweeter mentioned how much more reliable 600-year-old print technology is.

Well, that's true. Print is reliable. But consider: How many blind people can't access books that aren't also available in braille or audiobook editions? How many have trouble reading because of dyslexia? How many can't afford to buy a hardcover novel but don't live near a library?

A well-made ebook serves every one of these potential readers. There are dedicated text-to-speech e-readers for the blind. Some reading apps include dyslexic font options. Libraries have sizable ebook collections that can be borrowed and read on laptop, phone, tablet, or inexpensive e-reader.

The very nature of ebooks allows *all* readers to adjust size, font, screen orientation, and other features that make reading more accessible for everyone.

What some think of as display inconsistency when going from a small Kindle to a Kindle app on a large-screen PC is actually a feature. Yes, lines and pages don't always break nicely when the text reflows because we change font size. But the ability to change font size makes it easier for everyone to read a book in the most comfortable way.

We want everyone to read

- *when* they want to
- *where* they want to
- *how* they want to.

KEVIN CALLAHAN

For the editor

It's true: When working with ebooks, editors need to become familiar with technology and pay attention to unfamiliar details. But the challenges have a big payoff: More people than ever can read a book.

This booklet will introduce some new ideas and help you rethink old practices. I hope you come away from this excited to make books that are accessible to everyone and fun to read, and with the sense that ebooks are, after all, books.

Acknowledgments

I've had a fascinating career in editing, design, and production, but recent years have provided some of the most rewarding experiences. From my first exposure to making ebooks via Anne-Marie Concepción's courses on LinkedIn Learning/Lynda.com, to speaking at conferences she and David Blatner have hosted during CreativePro Week, to writing for *InDesign* magazine and speaking at ebookcraft in Toronto, to conducting webinars for the EFA and lecturing at Pace University, I've discovered the pleasure of learning by teaching.

Making books is always a joy, but my friends and colleagues make it a passion. Thanks for encouragement, inspiration, hope, skills, and tips to India Amos, Ron Bilodeau, Laura Brady, Karyn Browne, Liz Castro, Chad Chelius, Eric Christopher, Simon Collinson, Dave Cramer, Romain Deltour, Erica Gamet, Keith Gilbert, Sarah Hilderley, Jean Kaplansky, Naomi Kennedy, Nellie McKesson, Jude Naples, Susan Neuhaus, Iris Amelia O'Connor, Brian O'Leary, Jiminy Panoz, Sue Paré, Mike Rankin, Robin Seaman, Tzviya Siegman, Keith Snyder, Joshua Tallent, Jens Tröger, Cheryl Weissman, Colleen Cunningham Wnek, the BISG and members of the W3C EPUB3 Community Group, and so many more around the world. Our community is all about developing skills and coping with daily surprises with humor and determination. Great thanks to the EFA, and to Robin Martin and Cheryl Murphy for their invaluable editorial input. And, at the end of every day, my partner in life, Tim Morrow, gives me limitless love.

What's in this booklet?

I'm not going to spend any time discussing grammar or punctuation. Instead, I'm hoping to expand editors' concepts of what a book can be. Ebooks provide opportunities to make the text clearer and presentation useful in a unique way. Here are some things to look out for as you're reading.

Accessibility

We want to get books into the hands of everyone who wants to read them. So, ebooks include features that make it easier for folks to read.

Making ebooks accessible is another one-for-all responsibility. So, instead of a chapter on accessibility, I'm going to include tips and ideas throughout this booklet.

You'll also find ideas to make print books more accessible. Hopefully you'll look at editing—and reading in general—differently.

Some accessible features require extra expenditures of time and money on the part of the publisher, but many are easy to include, especially once you become accustomed to editing with them in mind.

Appendix B includes a refresher checklist of the ideas described throughout.

Cross-References

There are lots of cross-references here. If you're reading the print edition, keep in mind that all these cross-refs will be hyperlinked in the ebook. And if you're reading the ebook, go ahead and click that link! Use your device's **Back** feature to return to your starting point.

Terminology

I've included a glossary, but I want to establish some terms up front to avoid confusion. (Note: In the ebook edition, first uses of terms will be hyperlinked to the glossary for quick reference.)
Devices, e-readers, apps, software: These are all places where folks read their ebooks.
Readers, users: These are the folks doing the reading.
Ebook, EPUB, MOBI: An ebook is the item you have on your Kindle. An EPUB is the file that Apple has on their servers ready to send to someone purchasing an ebook. A MOBI is the Amazon version of the EPUB.
Ebook developer: The person who will take the print document and turn it into an EPUB, often (hopefully) someone with a good amount of design and book production experience. The developer can answer questions and let you know what's possible.

E-book? Ebook? eBook? EPUB? Epub? e-pub? ePub? ePUB?

I'm on the non-hyphenate team: I prefer ebook and EPUB. For one thing, it's cleaner. For another, the group that developed the standard has used EPUB for years. And anyway, who uses *e-mail* anymore? To avoid shouting, I'm making EPUB small caps.

I know this goes against *Chicago Manual of Style* and the *AP Stylebook*, which are sticking with the hyphen as of the 2020 editions. But many of us in the ebook-making community have taken the simpler route.

Ebooks and Editors

Here's the evolution of thinking at Penguin Random House. Lisa McCloy-Kelley, PRH Vice President, Director Ebook Product Development & Innovation, filled us in at the 2018 ebookcraft in Toronto. I'm sold!

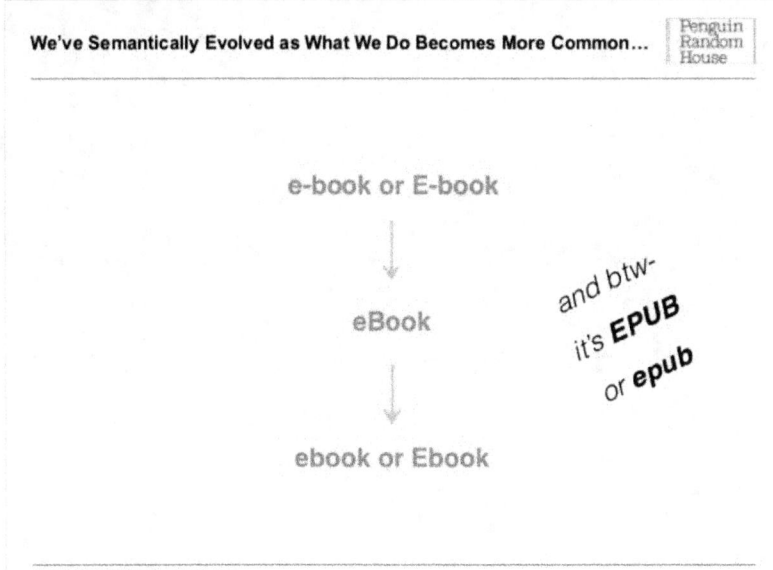

Chapter 1: What is an ebook?

An ebook is a book.

Like their print counterparts, ebooks contain text and images, but they have additional elements:

- accessible reading features, including customizable settings via font and size adjustments, contrast enhancements, and fonts for dyslexic readers
- embedded image descriptions to enhance text and caption information
- several layers and types of contents listings and book navigation
- inside-the-book hyperlinked cross-references, indexes, footnotes, and endnotes
- hyperlinks to external websites
- audio and video

Not every ebook has all these elements. There are differences in format and device capability. There are also differences in the amount of work a publisher will put into their ebooks, so while features like image descriptions may be helpful, they're not always a priority (they might be costly to create).

What are the ebook formats?

There are two formats: *reflowable* and *fixed layout* (sometimes called *fixed format*).

Reflowable ebooks are the most common. Think fiction, history, memoir, textbook. This is the most useful format because readers can take advantage of the most basic ebook feature: easy customization for reading ease and pleasure.

Text and images reflow after any font adjustment. Design usually resembles the print counterpart, but can't be a mirror image. In fact, it's not a good idea to try to exactly reproduce a print design in the ebook edition. They're different technologies! We'll discuss this more as we move through this booklet, and see Chapters 9 and 10 on design.

Fixed-layout ebooks are exact replicas of a print edition. This format is best used when text and artwork are tightly interwoven. A good use is for a children's picture book, where small bits of text are scattered around the page, often integrated with an illustration. One downside of fixed-layout ebooks is that fonts and font sizes are locked in; the reader can't adjust if the text is too small and usually has to pinch and zoom to read more easily. These ebooks don't contain some of the essential accessibility features, such as heading hierarchy or image descriptions that can be read aloud.

Is a PDF an ebook?

You may see some publishers refer to PDFs as ebooks. Retailers like Amazon and Apple don't sell PDFs as ebooks. You can upload a PDF to be converted to EPUB, but I don't recommend doing this. The conversion may create an inferior product, and authors and editors will have little chance to modify or correct any problems. In addition, PDFs are not like reflowable ebooks because they are not responsive to user display choice in the way reflowable ebooks are. This book will only address EPUB-based ebooks. (What exactly is an EPUB? Read Chapter 2.)

Where are ebooks read?

A seemingly infinite number of apps and reading systems display ebooks. Most readers either purchase ebooks from major retailers or download them from libraries. Here are the most common options:

- **Kindle (from Amazon):** color tablets, e-ink readers; apps for iOS, Android, Macintosh, PC
- **Apple Books** (previously *iBooks*): iPads, iPhones; Macintosh desktop app
- **Kobo:** color tablets, e-ink readers; apps for iOS, Android, Mac, PC
- **Nook:** color tablets, e-ink readers; apps for iOS, Android, Mac, PC
- **Google Play:** iOS and Android apps; Chrome browser plugin
- **Libraries:** via services such as OverDrive, Hoopla, and Freading, often using the **Libby** reading app or **Adobe Digital Editions**.
- **Many other e-readers** are available from companies like PocketBook and Boox.

E-ink devices

E-ink refers to the screen technology; these devices mimic printed paper, and are glare-free. Until the last couple of years, they have been grayscale only (black and white), but color e-ink readers are coming to market at an increasing pace. So far, though, the big e-ink device makers (Amazon and Kobo, for example), only offer grayscale versions. Customers can download a book from the Kindle store to their color Fire tablet and later switch to reading on the e-ink Voyage e-reader. We can't predict what a reader will do once they've bought a book, so we have to account for all possibilities. For example, designers, editors, and ebook developers have to be careful about color for those e-ink users. We'll touch on color issues in Chapter 10.

E-reader consistency

Reading an ebook on a color Kindle Fire and on an e-ink Kindle Voyage provides the same reading experience, right?

No.

Device makers bring different rendering capacities to each platform. Drop caps display differently on different devices, for example. This is important to remember when doing QA (quality assurance; read more in Chapter 11).

Software in devices and apps is frequently updated to include more display capabilities. But not all owners update their devices, so developers have to keep in mind older, less sophisticated reading apps.

It's unpredictable!

Device uncertainty is largely a problem for ebook developers not editors. It leads to frustration and annoying problems when creating ebooks. Editors can help by ensuring the text is well structured and consistent.

Who makes ebooks?

- ebook developers
- book designers
- web developers

with important assists from

- authors
- developmental editors
- copyeditors
- production editors
- proofreaders
- indexers
- designers
- production artists

Remember: One for all, and all for one!

What's the editor's role?

If a manuscript you're editing will be published as an ebook (a very likely situation), you'll do a more complete and universally usable edit if you know what to expect in that channel. So, read ebooks.

Even if you don't want to buy ebooks, you can open an account with Amazon or Kobo and download their apps onto your smartphone or laptop. Apple devices come with Books already installed, so there's no downloading needed. Once you have an app or two installed, download book samples from the Kindle store or borrow books from your local library.

Chapter 2: What is EPUB?

Ebook publishing has its own jargon. The glossary provides a thorough list of terms and their meanings, but for now, we only need to focus on just one term: EPUB.

The basic ebook file type is EPUB (like Word's *.docx* or Acrobat's *.pdf*), and serves as the suffix at the end of a file name: *book.epub*. Ebook developers create the EPUB and send it to the editor for quality assurance.

All retailers and distributors accept EPUBs. Until late 2020, Amazon allowed upload of its own file format for use on the Kindle: the KF8/mobi (*book.mobi*; KF8 refers to the version currently in use). As of this writing, Kindle Direct Publishing (Amazon's book publishing center) prefers that authors and publishers provide the EPUB.

The mobi was a descendant of sorts of EPUB. It was converted from the final EPUB using Kindle Previewer or KindleGen, Amazon-created tools.

Anyone reviewing an EPUB can look at it on Kindle Previewer for an approximation of how it'll appear and function on several Kindle platforms. You can also export a mobi from Kindle Previewer and sideload the file onto your own Kindle device for checking.

Once you upload an EPUB to Kindle Direct, Amazon rips it apart and reassembles it to create the download-ready product.

EPUB2 or EPUB3?

The current standard is EPUB3. It includes support for semantic (or *grammatical*) XHTML markup, robust navigation (numerous tables of contents, and other get-around-the-book aids), language tagging, and multimedia possibilities. It allows for a rich reading experience and is the basis for whatever the future holds for ebook reading.

Some retailers were slow to adopt EPUB3, but all accept it now. Because of that slow acceptance, some publishers still create EPUB2 files and some distributors wrongly ask for it. Those old-format files can still be read on every device and app, but some sellers are demanding that backlist EPUB2 files be upgraded to EPUB3. So, there's no reason to not start out with EPUB3.

Older devices will read EPUB3 files as long as backwards-compatible navigation files are included, which is the ebook developer's responsibility.

What are EPUB standards?

Ebooks are built on the same technology as websites, so the World Wide Web Consortium (the W3C) controls the standards for both platforms.

The W3C's Publishing Working Group (Publishing@W3C) is dedicated to expanding the possibilities of digital book publishing. It is composed of book-industry professionals from publishers around the world, along with reading-system developers.

The Publishing Working Group surveys publishing needs for all kinds of publishing—textbook, trade, journal, academic—and develops methodologies for presenting information using web-based standards. There are some features of the standard that few reading systems take advantage of. For example, the Page List is a hyperlinked listing of the print edition's page numbers. Each link takes the reader to the spot in the ebook that corresponds to that place in the print edition. So, two people in a book club can be on the same page. That is, it works if the app displays the Page List. Many don't as of this writing.

What's in an EPUB?

Some of the same building blocks in EPUBs are in websites.

- XHTML documents: where the content lives. Text is tagged with structure and formatting instructions: chapter title; epigraph; italic or bold. Each XHTML file typically contains a chapter, so a twenty-chapter book will have at least twenty XHTML files. There will also be files for the cover and the table of contents.
- The CSS (cascading style sheet) defines the design. Is the chapter title centered or flush left, black or purple? Is the epigraph a different font from the rest of the text? All those choices are detailed in the CSS.
- Images, including illustrations, photographs, and sometimes tables.
- Fonts, which can often be embedded. Note that designers or developers need to ensure that embedded fonts are licensed specifically for ebook use.
- Media (audio, video), which can be included but are played only on some e-reading devices or apps.
- Some EPUB-specific files that tell a device: *This is an ebook.*

What about metadata?

Metadata exists in two locations: within the ebook and on a retailer's website.

- The metadata within an ebook is not discoverable; that is, a web search by someone looking for carrot recipes will not have access to the metadata in a recipe ebook.
- If the right keywords are included in the metadata in the Kindle or Apple store, that book will be found on the retailers' sites.
- The ebook developer can embed metadata in an ebook, and some are required (book title, author, accessibility features). But whoever uploads the book to Amazon or Kobo must submit the more-thorough, discovery-oriented metadata on upload.

Conversion or adaptation?

Should an ebook be a replica, in content or design, of the print edition? Not at all. Print and EPUB are completely different technologies with different requirements. That's why I never say that I *convert* a print mechanical into an ebook. Instead, I *adapt* design and content to match the requirements of digital publishing.

What's the editor's role?

It's helpful to know how an ebook is made and what pieces it contains. Most editors won't need to dig into the files, though. You most likely won't have to do your editing in the XHTML.

Chapter 3:
What's the workflow?

Here's a simple process for making an ebook:

- Write in MS Word (or Google Docs or other word processor).
- Edit. Assign paragraph and—especially—character styles.
- Flow text document into InDesign for page layout. As part of import, map (or sync up) Word styles to corresponding styles in the InDesign document.
- Design and lay out print edition.
- Export from InDesign to EPUB. As part of export procedure, map paragraph, character, and object styles from InDesign to the EPUB's CSS.
- Open EPUB and modify, clean up, add features.
- Validate the EPUB and upload to distributors and online stores.

Until the fourth bullet point, it's pretty familiar. And only the first two are strictly editorial, but what an editor does in those steps has a big impact on all the tasks that follow.

Note that this workflow includes book layout in InDesign, which is the tool of choice for most book designers. However, many publishers have their own typesetting systems that don't include InDesign or don't use InDesign in the same way. Others use QuarkXpress. But however the book is laid out, the manuscript has to be organized and tagged properly.

What are some editorial concerns?

Once you've read some ebooks and have become familiar with how reading systems work, think about your role in making them. Here are some ideas to kick around in early editing stages:

- Which features of the print edition will need reworking? Tables, images, sidebars, pullquotes, and print-specific design elements may need to be rethought or redesigned for the ebook edition.
- What content can be added? Extended bibliography? Study questions? Ebooks have no real length limitations, so anything can be considered.
- What extra features—audio, video—are desired and possible? How do target reading apps display them?
- How are all these elements made fully accessible?
- How can ebook design reflect the print design?
- Do you have an ebook-making team?
- What editorial and production workflows are in place?

Write the manuscript.

Now's the chance to get the author involved. If she is comfortable with Word's Style Pane, encourage her to use it or coach her if necessary. A well-tagged manuscript will make the editor's job a lot easier.

Edit and apply styles in Word.

Does the author include elements that won't fit in print but can improve the ebook? Make a note. Are there design features that you'll want for print, but know won't work in ebooks? Note them, too. Are there lots of cross-references? Make sure they're clearly marked.

As you decide on content, think about structure. Apply Word's paragraph and character styles. Are the heading hierarchies logical? (Make sure Heading 1 leads to Heading 2 and on to Heading 3; don't skip Heading 2.) Are paragraph styles applied to all paragraphs, including heads, text, extracts, lists?

When applying paragraph styles, it's usually fine to leave the appearance basic. There's no need to exactly specify how much space surround a subhead, for example. That's usually up to the book designer and typesetter (or page compositor).

Are bold and italic terms and small caps marked with character style tags? (Read more about italics and bold in Chapter 7.)

Whatever styles you apply, please make sure all attributes are included in the style definition, and are not manually applied. If an element does not have a style applied, it may well not be noticed by the designer or typesetter or ebook developer and disappear.

Once you've applied styles, you won't need all those extra paragraph returns and tabs that authors like to use to format their manuscripts. Deleting these extraneous items will deliver a cleaner manuscript to the designers and developers.

Are there exceptions to the dominant language? If the book is in American English with a smattering of French or Greek or Swedish, tag those words with a unique character style. This will export to the EPUB with the correct language tag, which will be helpful to text-to-speech renditions. Make sure to tell the ebook developer about these language tags.

Are there lots of images, tables, and sidebars, and are they called out in the manuscript, either within the text or as comments? The ebook developer usually places these items directly after the paragraph in which they're called out, so marking their ideal location is important for work down the line.

Being rigorous about applying styles will make the designer's, typesetter's, and developer's job easier and more predictable because everything requires a style. You're passing an accurate manuscript to design and production that will result in fewer errors down the line, including at the ebook-creation stage.

Flow the manuscript and lay out pages.

The designer (or typesetter; many publishing houses break these functions into separate jobs) flows the manuscript and composes pages. The editor's efforts at organizing and tagging content will pay off here, because of the magic of mapping.

Map styles from Word to InDesign.
Part of the print typesetting process—at the very start, in fact—is correlating Word's paragraph and character styles to the typesetter's style sheets. They could have similar names (Heading 1 maps to h1), but they don't have to. The typesetter will make sure that what you tagged as Heading 1 stays first level in the designed pages. Just be consistent and logical in style tagging.

We're almost ready to export to EPUB.

When the print edition has been finalized, it's ready to export to EPUB. InDesign has a tool that will do this. The ebook developer will most likely have to do some housekeeping, particularly if he didn't do the typesetting. Images and sidebars may need to be anchored near their text reference; cross-references and endnotes may need to be hyperlinked.

Map styles from InDesign to CSS.
Remember those styles you applied in Word? They still exist, but now in the typeset mechanical. The next step includes mapping those InDesign styles to CSS styles for use in the EPUB.

Ebooks and Editors

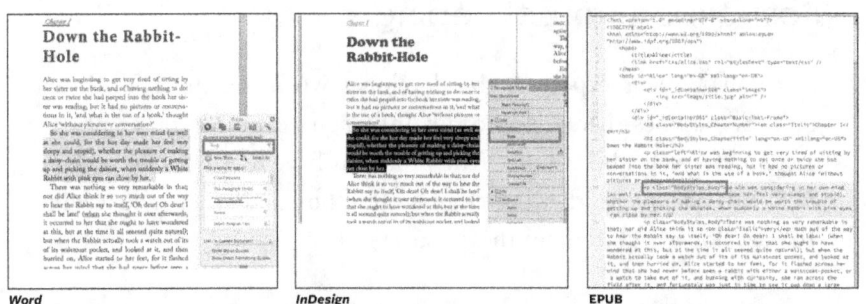

Word   InDesign   EPUB

In this image, note how the paragraph is assigned the paragraph style Body in all three locations. Tagging during editing smooths the process all the way to the end.

The table below presents a style sheet comparison. XHTML markup is the only set of names that cannot be changed. Word style names are built in to any document you open, and you can change them to match the final XHTML names if you want. InDesign names are up to the user.

Word style	InDesign style	HTML markup
Heading 1	h1	\<h1\>
Headings 2 . . . 6	h2 . . . h6	\<h2\> . . . \<h6\>
Normal	Text	\<p\>
Body Text	Text	\<p\>
Body Text 2 (no para indent)	left	\<p class="left"\>
Body Text 3 (chapter opener para)	co	\<p class="co"\>
Block Text	blockquote	\<blockquote\>
List Number	num-list	\<nl\>
List Bullet	bull-list	\<ul\>

Post-export housekeeping

Once an EPUB is exported, the developer works on the HTML and CSS. Sometimes there's lots to do; InDesign's export is never pristine despite everyone's best efforts. The problems have mostly to do with the XHTML documents that InDesign creates, not content or style tagging. The developer will iron out these issues.

Add metadata

Metadata can be added at any time from InDesign layout to post-export EPUB work. The developer can usually get most of the information from the book itself:

- book title, subtitle, author(s), author title(s), publisher, edition, revision number
- book description, keywords
- ebook ISBN, copyright status and holder, copyright information email, publisher, rights
- accessibility features

All these entries should be in the ebook. They must also be provided to the retailer/distributor on file upload.

Validate the EPUB

Once the EPUB has been proofed and is final, the developer will validate it. That is, he will use an application (a commonly used app is EPUB-Checker) that examines the file's structure and completeness. If any of the markup is incorrect, if pieces are missing, or if the navigation has problems, the EPUB will fail validation and need to be fixed. Retailers will turn down a file that fails validation.

Chances are the developer will validate more than once during the workflow, but it's important to know that the file is valid before upload to a distributor.

A separate accessibility validation is available: ACE by Daisy. It checks head hierarchy, the presence of ALT tags for images, and metadata. This level of validation is not necessary for upload to Amazon, Apple, or any other distributor, but it's an important check that the EPUB is well made and accessible.

The Resources in Appendix D lists some validation tools.

What if all I have is a PDF or OCR scan?

Lots of backlist titles exist only in PDF or as scanned text. They are often brought back to life only for ebook publication, but sometimes they are going back into print as well.

The first step is to extract text from either the PDF or a scan. Then, proofreading is essential. Lots of errors can crop up.

If there's a print edition in addition to the ebook, design and layout are next. But if the book is going straight to ebook, then the usual manuscript-prep steps need to be taken. There can still be an intermediate step between the optical character recognition (OCR) manuscript and EPUB creation (like a set of galleys), but all heads and other elements need to be tagged, all images anchored, and so on.

Cleaning up a PDF or OCR scan manuscript

There's a bit of work to do to ensure the file is ready to convert to EPUB. Here's a brief rundown of the process:

- Extract text and images from PDF.
- If the PDF was created directly from applications like Word or InDesign, it should be pretty clean and without too many errors.
- If the PDF was created by optical character recognition software, where a physical book is scanned, look out for scanning errors (like capital *S* rendered as *5,* or backwards-facing quote marks). The fidelity of the scan depends on the quality of the print source.

- Run find/change to fix OCR issues, or
- Make a list and give to typesetter or ebook developer.
- In Word, edit and apply styles.
- Flow into InDesign.
- Map styles from Word.
- Apply styles as needed.
- If a print edition is not part of the plan, create galleys with all text styled.
- Proofread to make sure text is ebook- or print-ready.
- Create EPUB.

Chapter 4: Semantic content matters

Like print, digital publishing draws a bright line between the text and how that content is presented. But in ebook building, visual cues and graphic devices are less important than hierarchy and function. What's a head? What's a sidebar? Is this word boldfaced because it needs to be emphasized, or is it just a design choice?

Imagine a print book designed as plain text: Every word is set in 12 pt Times Roman, with no paragraph indents or space between paragraphs. It would be difficult, annoying, and maybe impossible to decipher. Even if we assigned heading styles, but they all shared the same attributes of 12 pt Times Roman, it would be hard to figure out.

In an ebook, so long as the chapter head is tagged *h1*, the first subhead *h2*, and a pullquote *aside*, a reading system can make sense of it. The elements h1, h2, and aside have specific meanings and functions. This isn't much different from the manuscript markup copy editors often do. Terms might vary, but the concepts are the same.

Using [assistive technology (AT)](), the heading hierarchy can be navigated (by reading all the h1 entries, for example), and the asides can be ignored if the reader wants to get through the main text without being interrupted by sidebars or pullquotes.

An ebook's style sheet (the CSS) defines appearance. In case the CSS somehow disappears, every reading system has built-in display instructions that differentiate the presentation between h1 and h2. So ebooks have built-in capacity to display a book that visually and semantically makes sense.

EPUB semantics

Semantics is a term you'll hear a lot when talking to ebook developers. It's another word for *grammar*, I guess you could say, or structure. Here's a partial list of semantic markup:

- *section* marks major divisions of text (chapters; each level of heading within a chapter)
- *h1–h6* are the heading levels available. Every head—from the title (h1) on the title page to chapter titles to subheads—must have a heading level
- *aside* indicates sidebars, pull quotes, footnotes; any ancillary material that can be read after the main narrative
- *blockquote* marks excerpts
- *figure* and *figcaption* label images and their captions

Apply structure

Imagine a novel broken up into parts, with a few chapters within each. The part number and title are the first-level heads (Heading 1 in Word; h1 in InDesign and XHTML). The chapter number and title are second level (Heading 2; h2). Each subsequent heading will move down the list (Heading 3–6 in Word; h3–h6 in the XHTML).

It might be tempting to label the part number as Heading 1 and the part title as Heading 2. It doesn't matter in print. I can design the number and name any way I want; the visual hierarchy helps comprehension.

But for AT to work as intended (reviewing a book by having all h1 heads read aloud, for example), those two items—number and title—are of equal importance, so both are Heading 1. Now the question is: How should the editor present the part number and name? Look at Options A and B:

Option A
Part One
The Dog Arrives

Option B
Part One: The Dog Arrives

If the developer leaves each line as a discrete heading (Option A), someone listening to all the first-level headings will hear a staccato rendition of the part opener: the part number first, then the title when advancing to the next h1.

In Option B, AT will read the opener as a single unit. The next h1 will be Part Two.

The print designer will arrange the heading any way she chooses. Sometimes a designer will decide to put the number on the opposite side of the page—or spread—from the title, or running up the page sideways, making this assembly difficult to maintain in the print layout. All an editor can do is present the manuscript using this method and call it to the designer's attention, and eventually to the ebook developer's attention.

Navigation

Navigation is just another word for getting around an ebook. Specs for EPUB3 call for these hyperlinked lists:

- embedded navigational table of contents (viewed in a reading system's built-in navigation panel and referred to as the ncx for EPUB2 and toc.xhtml for EPUB3)
- landmarks (appears as part of the reading system's navigation)
- page list (appears as part of the reading system's navigation)

And, you can add extra navigation tools to help the reader get around the ebook:

- internal table of contents (placed somewhere in the frontmatter)
- lists of illustrations, photographs, tables, etc.

Tables of contents, illustrations, tables

Retailers require an embedded table of contents that is part of every reading system's interface. Even if the print edition doesn't have a table of contents, one must be embedded in the ebook. It can be as simple or exhaustive as you want.

You can reproduce that embedded listing—or an edited version—in the body of the book, as part of the *frontmatter*. Other lists are great additions. Readers will be more informed of a book's contents as they view a retailer's sample (*Look Inside* on Amazon, for instance), and the listings will provide extra help navigating the book.

The good news for the editor is that the typesetter or ebook developer can create all these lists and navigation aids. Just decide how deep into the heading levels you want to go, and which elements you want listed.

Landmarks

Landmarks are an EPUB3 accessibility tool.

- They mark major elements: start of content, part, chapter, preface, backmatter, bibliography, index.
- They help users of assistive technology navigate.
- The *bodymatter* landmark tells the device or app where to start reading the first time it is opened on that device. It can be set to the title page or another part of the front matter. Some reading systems override this and start at the first chapter.

Page List

The Page List is another part of the EPUB3 toolkit. It tags locations in the ebook to correspond to print page locations. If members of a book group are using both the hardcover and the ebook editions, all can find *page 23* easily.

The ebook developer creates the Page List.

Note that Landmarks and Page List are not displayed across the ebook-reading universe. Some software, like Adobe Digital Editions, display both. Most reading systems don't. But they are worthwhile to include as part of the EPUB3 spec.

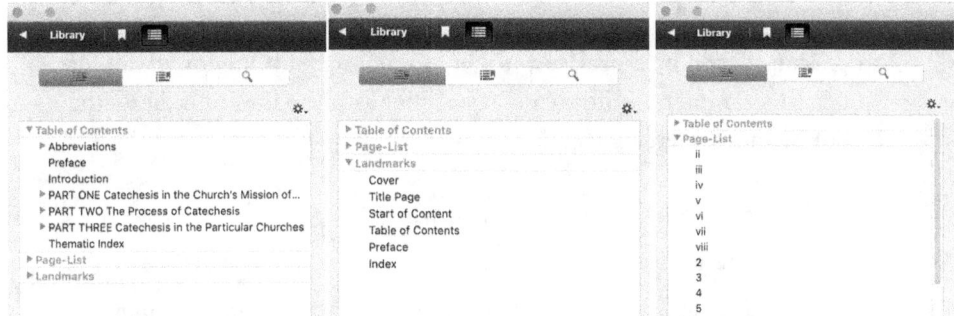

Above is a view of navigation panel in Adobe Digital Editions. On the left, the table of contents. Each Part has a flyout menu that contains more levels of headings. In the center, a simple list of landmarks. On the right, the beginning of the Page List.

ARIA roles and epub:type

ARIA roles are used in web development as well as in ebooks, but epub:type applies only to book publishing. They define content and are even more specific than landmarks. There are roles and epub:types for the Introduction, Preface, and Index, parts, chapters, and other high-level elements, but there are also ARIA designations for granular elements like glossary terms and bibliographic entries.

As technology improves, ARIA roles will help readers using assistive technology navigate a book, but epub:type has no real function in AT. So, while epub:type is not necessary for AT, it is still helpful:

- epub:type is available in InDesign for a typesetter or developer to assign, if identified by an editor.
- ARIA roles must be added by the developer in the XHTML, with editorial guidance.

- If epub:type is assigned within InDesign, the developer can add corresponding ARIA roles post-export.
- The Resources section in Appendix D has a link to the Daisy Knowledge Base, which provides a full listing of ARIA roles.

The typesetter and ebook developer will assign these labels, but the editor can help by noting them in the manuscript. Some are obvious, but on occasion there'll be a preface with an unusual name (One for all, all for one!, for example). In those cases it's helpful for the editor to indicate the book element. Here's a sample.

Landmark	epub:type	ARIA role
Preface	preface	doc-preface
Introduction	introduction	doc-introduction
Table of contents	nav	doc-toc
part	part	doc-part
chapter	chapter	doc-chapter
Index	Index	doc-index

Chapter 5: Expand content

When a hardcover novel is reissued in paperback, no one questions if it's the same novel. When it's recorded as an audiobook, we take for granted that the narration is faithful to the print document.

Well, an ebook is the same. It transfers content from one format to another. It may have some features added, but the heart of the book is as faithful to the print edition as are paperbacks.

This chapter provides more adaptation ideas: what you can add or change to enhance the book in its EPUB incarnation.

Expand content

There's no page count in ebooks, so feel free to add features. There are almost as many options to enhance a book as there are books.

Beef up references; add extra navigation, as discussed in Chapter 4; use color images; add image galleries; hyperlink cross-references and glossary entries.

When you begin editing, gather ideas of what you'd find helpful in the ebook edition. Talk to ebook developers to see what you can include and how.

Modify the book

We're accustomed to the order of elements in a print book. But what makes sense for the ebook edition? Here are some ideas. Remember that reading an ebook is different from print. It's not as easy as sticking your thumb in your place and flipping through the pages (you need to use the reading system's bookmark instead).

Front matter

- Think about people shopping in the Kindle Store (better yet, browse the Kindle store yourself). They'll click on the *Look Inside* feature. What's important to see in that sample?
- Does the copyright page have information that will make anyone want to buy the book? If not, move it to the end (and add a link to the table of contents).
- What can be added to the front matter? Maybe include *How to Read this Ebook*, a cast of characters, or listing of acronyms and their definitions.

Body of the book

- What text modifications need to be made (rewriting link language; rephrasing cross-references)?
- How should complex tables be configured (see Chapter 8 for more)?
- Is it possible to keep images with their captions (see Chapter 10 for more)?
- How does linking work (cross-references, indexing, to external websites)?

Back matter

- What can be added? An index of ingredients to go with the recipe index?
- What can be enhanced/made interactive with links? Bibliographies should include hyperlinks where possible. (But watch out for Link Rot, discussed in Chapter 6.)
- Can the index be created so references go directly to the term within the book or will it be linked at the page level?

How are language variations handled?

Here is another accessibility feature. If a book's main language is American English but includes another language, text-to-speech needs to be told that there is an exception to the rule.

Apply unique character styles to each word in a second or third language and alert the typesetter and ebook developer. Alternatively, provide a list of words so the typesetter can do global searches and apply the tags.

Are lists problematic?

Multilevel lists—numbered or bulleted—can be difficult to read on narrow devices like a phone. Each new level indents from the left, so the more levels, the more indentation, until you end up with an indent level that may allow only a single, short word.

Rewrite multilevel lists, if possible. Review the ebook on a phone.

Edit the copyright page

Whether or not you move the copyright page to the back, there are some modifications needed:

- delete printing history and font information (unless embedding)
- include the ebook developer in credits
- add ebook-only copyright language and e-ISBN

Chapter 6: Hyperlinks

A valuable feature of ebook reading is the ability to move around a book at the tap of a hyperlink. Links can take you

- from the table of contents to a subhead deep in a chapter
- from one chapter to another
- from a term to its glossary definition
- to discussions within a chapter
- to external websites

NOTE: Reading devices usually include a built-in, on-screen *back to page XX* link so you can return to your starting point.

Provide a global instruction to the ebook developer to create these links. For example, apply a dedicated character style to cross-reference text so it's easily found. Similarly, assign a character style to terms defined in the glossary.

The typesetter or ebook developer will create the links, but it is helpful to have them pointed out. If you've created any of these links in the manuscript, alert the print and ebook teams following you.

Footnotes and endnotes

Footnotes and endnotes generally include a superscripted number or symbol that links directly to the note. These should be clear in the manuscript through the application of character styles.

- Footnotes should be round trip (from the text to the note and back again). The developer takes care of this functionality.
- Since there may be multiple text references to a single endnote, round-tripping isn't always possible. So, a second and third text reference to a given endnote can only go to that note, relying on the reading system's *back to XXX* function.

Use Word's tools to create footnotes. These transfer to InDesign, and then on to EPUB.

Tell the typesetter or ebook developer when there are multiple callouts to a single endnote. Knowing the complexity of the manuscript will smooth the process of creating linked notes.

External websites

External hyperlinks must be live (clickable) and current. Keep in mind that

- links get old and stop working (commonly called *link rot*)
- the user might not have an internet connection while reading
- a reader who follows an external link may not come back to the book
- too many links to outside (or in-the-book) destinations can be distracting

Ascertaining hyperlink viability is an editorial task. An ebook developer can test if a link is active, but can't say if it goes to the correct site. That responsibility rests with the author and editor. Here are a couple of ideas to wrangle links:

- Maintain all external hyperlinks on the author's or publisher's website, and point there from the ebook. This will bring readers to the website (where they can see other material on offer) and is easier to maintain.
- Subscribe to a link archiving service such as the Wayback Machine from Internet Archive (see Resources in Appendix D for a link).

Hyperlink language

Links should obviously be links, both visually—with underscores and in color—and in language used.

- Rewrite a text reference to make clear where the link goes: Click here for more information about cooking eggs.
- Try to avoid one-word links, particularly small words. Make link phrases long enough to be visible and clickable. Clicking on or tapping a short word on a phone can be tough.
- Make link language consistent. For example, choose one of these approaches:
  - Read about Legal Priorities *or*
  - Read the Legal Priorities discussion.
- Tell the ebook developer that links should have both a color that is different from the text and an underscore. Ebooks with lots of underscored links can be distracting, so it may be tempting to use only a color to indicate a hyperlink. But not everyone reads on a color e-reading device and some folks have poor color perception. Using both color and underscore is sure to emphasize that there is a link. Kindle reading software adds underscores even if you don't want them, and Apple applies its own color instead of one the developer asks for in the CSS.

Link it up, but don't distract

It might be tempting to provide lots of links to move readers around the book and out to the internet, but be judicious. Think about blogs you've read. Some writers like to embed links for many references. They can become distracting and annoying. On occasion I'll even click one and never come back.

Index everything

By now you get the idea that ebooks present a chance for an expansive experience. Here are some ideas for extra hyperlinked indexes:

- recipes
- ingredients
- tools
- characters
- place names
- any kind of specialized index you can think of

What's the editor's role?

- Plan ahead. Build indexing into the production schedule to allow for release of the final typeset file to the indexer before printer files are needed.
- Engage indexers who work within the typesetting program, like InDesign, so they can create a fully linkable index.
- Never publish an ebook index with unlinked page references. Linking to the term itself after the index is written may be daunting and expensive. At the least, ask the developer to link a page reference to the text at the beginning of the page indicated. LiveIndex is an InDesign script that will do that (see Resources for more information).

Chapter 7:
Which italic is this?

The semantics of EPUB contain multitudes of italic and bold choices. It's not enough to highlight a word and click the *I* or **B** button in Word. The editor must distinguish among three types of italic and two bolds. It's not hard; most italics will be just italic. But the differences are important.

These distinctions are semantic (grammatical) necessities and they give text-to-speech software instructions in how to voice a word.

How many italics and bolds are there?

There are three types of italic and two of bold in semantic markup.

<i>, <em>, and <cite>
- <i> Eloise planned to travel on the *Queen Mary*. (Marks change in tone or voice; technical, foreign word; does not need to be stressed by text-to-speech.)
- <em> She *had* to sail on Monday. (For vocal emphasis in text-to-speech.)
- <cite> As they left port, she thought of the movie *Titanic*. (Semantic markup of citation.)

\<b\> and \<strong\>

- \<b\> **Ship's log:** The captain of the ship maintains the log. (Bold indicates semantic meaning; here, a keyword; not vocally stressed.)
- \<strong\> **Warning:** The ship will capsize at 0200 hours. (Bold is meant to be vocally emphasized.)

Use existing character styles or create new ones in your word-processing program, then tell the typesetter/ebook developer which kind of italic is which and they'll apply the correct semantic XHTML markup. You can even name the character styles using the italic and bold types. That'll make everything clearer.

Chapter 8: Tables, math, and science

Because of the variety of screen sizes and reading app functionality, complex tables are difficult to configure for digital publishing.

Simple tables are usually acceptable, but tables with straddle heads and multicolumn rows and bulleted lists can be impossible to read on many reading platforms. What are the options?

- Rewrite into narrative form (bold head with paragraphs beneath).
- Break into smaller pieces.
- As a very last resort, include as image, but provide fallbacks:
- Include a live-text table on a website, with a link from the ebook.
- Include ALT text describing the main points.

What's the editor's role?

Early in manuscript preparation, devise your approach for complicated tables in the ebook. Can a table be presented another way in both print and ebook, as a list, for example? Recognizing that tables often need to take up a lot less space in print than in ebook (print often has page-count limitations, while ebooks don't), think about this solution.

Book element	epub:type	ARIA role
Preface	preface	doc-preface
Introduction	introduction	doc-introduction

This can become:

- **Book element:** Preface
- **epub:type:** preface
- **ARIA role:** doc-preface

The developer's role

Good table construction is not just an issue of its appearance on screen, but also has accessibility ramifications. Assistive technology can move through a table reading column heads, for example. Ask your developer to make sure the XHTML is compliant with accessibility needs.

Math and science notation

Equations also present a problem in ebooks. The EPUB3 spec calls for using MathML in websites and ebooks. However, support in reading systems is inconsistent. Publishers that produce books heavy in notation will have their own systems for addressing this issue. The temptation is to render equations as images, but as discussed in Chapter 10, ALT text needs to be included.

The Resources section in Appendix D includes links to discussions of MathML.

Chapter 9: Ebook design

Ebook design is different enough from print design to have its own set of best practices. The first shock for folks coming from print is that there's no foolproof way to assert control over how text is displayed.

We may decide that chapter titles should be in sans-serif bold, centered, with three lines of space until the first paragraph. But a reading system may decide that all chapter titles should be flush left. Another reading system might only want to use a serif font for heads. So much for the designer!

The CSS author

The CSS is the style document in an EPUB that defines the look of the ebook.

InDesign will write a CSS document on export to EPUB. It translates the print design into something that might work well in an e-reader. To maintain as much control as possible, I use my own CSS. It makes allowances for what I know works and doesn't work across the board. InDesign's CSS doesn't have that experience built in, so it may write a CSS file that won't create a reliable reading experience.

When I export an InDesign document to EPUB, I first make sure all semantics are in order (all the h1, h2, aside, blockquote tags are assigned), then attach my own CSS document. I just need to modify the attributes in the CSS to create the ebook design.

While we're talking about design . . .

Some best practices for ebook design differ from the print norm. Most aim to make reading easier.

- **Flush left, rag right.** Reading-system hyphenation dictionaries can create hard-to-read word spacing in justified text.
- **Hyphenate text.**
  - But don't hyphenate heads. Long words like *Acknowledgments* can run off the screen; the developer should make allowances.
  - Ask your developer to include hyphenation guidance in the CSS. It is possible to specify two hyphens in a row, for example (but not all reading systems will pay attention).
- **Make spacing both a little tighter and a little more generous.**
  - Decrease space around heads (remove space between heads and text, for example). This helps keep heads with the following paragraph.
  - Use generous line spacing (leading, called *line-height* in CSS) for on-screen readability. Many ebooks have very tight line-height, which makes for unpleasant reading.
- **Winnow font varieties** used in the print design (e.g., if the designer uses many variations of a typeface, like Thin, Book, Regular, Medium, Semibold, Bold, Black). Subtle distinctions can be lost on screen.
- **Modify hyperlinked superscript and subscript numbers.** Footnote and endnote text references can be hard to see and tap on screen when the numbers are too small. Try larger or bold numbers, or enclose in parens: *(1) (2)*. If there are lots of sub- or superscripts, ask the developer to experiment with presentation so they are obvious and clickable but don't overwhelm the content.
- **Add ornaments for narrative breaks.** If the print edition indicates a pause by a line space followed by a flush-left paragraph,

consider adding a small row of asterisks, bullets, or something more ornamental in the ebook. Readers may not see the line space if it lands at the bottom of a screen and won't always notice the flush-left paragraph at the top of the next screen.

- **Avoid text wrapped around images.** Text wrap is inconsistently displayed across the vast universe of e-reading software. Sometimes it's downright ugly (too much space around an image or not enough space or space in the wrong places). So, it's best to just place items that were text wrapped in print into the text flow between paragraphs.
- **Avoid drop caps.** These suffer from the same fate as text wraps: poor, inconsistent rendering across devices and apps. It's better to bold the first word or sentence, or just leave unadorned.

Chapter 10: Color and images

You can upgrade your editorial toolkit if you understand a few things about ebooks, images, and color. These are usually design issues, but editorial input can ensure accessibility and easy reading.

Color and contrast

Not everyone reading your book will be using a color device, and some readers may be color blind. With that in mind,

- Don't use color to define content (avoid *"see the numbers in the red column"*).
- Rewrite as needed (*see the numbers under the column head* Participation Rate).
- Check color combinations. Some combinations are difficult or impossible to see for some readers, and e-ink devices may not display legibly.
- Check contrast. Everything must be legible on e-ink devices. Look at the image to the right. Pale gray text on a slightly darker gray background may be elegant in print, but could well disappear on an e-ink screen (and may not be legible for some readers in print, either).

There are websites that grade color use for the web and ebooks; see the link in the Resources in Appendix D. This is applicable to print accessibility as well. Try to remember all readers when editing a manuscript or reviewing page proofs.

Images

One question comes up almost every time a book includes images and captions: *Why won't my caption stay on the same page as my photo?* The answer:

- Ebooks reflow, so all text gets bigger or smaller.
- Captions take more or less space depending on font size.
- So, captions will sometimes spill over to the next screen.

One terrible but popular solution is to capture the image and caption as an image. Why terrible? The caption becomes unavailable to screen readers, so anyone using assistive technology will miss out on both the image and caption. Yes, it can be included as ALT text (see below), but what if you need the ALT text to describe the image itself? And, think about viewing the caption-as-image on a phone screen; it would be pretty small and probably difficult to read by pinching and zooming.

Here are potential solutions. Depending on the book, is it feasible to

- Force every image/caption combination to begin on a new screen? No matter how text reflows, the photo will always appear at the top of a screen, with the caption beginning below.
  - The downside: You might have a very short page leading into the image. That can be distracting or confusing for readers.
- Break the image and caption apart on purpose?
- Place the caption at the top of the next screen and lead it off with a parenthetical (Previous page).
- Place the caption before the image and add (Following image).

What's the editor's role?

- Ask the ebook developer to use settings that make vertical images resize depending on screen size. An image can be coded to use a portion of the screen's vertical space, leaving the rest to contain the caption.
- Acknowledge imperfection. Even if you force an image to the top of a screen with the caption right below, the caption might still break across pages depending on its length and the user's font size.

ALT Text

Writing ALT (alternative) text is an important editorial task in building accessibility. It is embedded in the image in the XHTML and is read out by assistive technology. It ensures that anyone using text-to-speech has access to all the important information in an image.

For readers using text-to-speech to understand an image, it must be fully described:

- in the surrounding text
- in the caption

If there's still more information than is described in text and caption, then it must be included

- in ALT text

ALT text should be about 140 characters. Longer descriptions can be embedded in the ebook, linked from the caption.

Tag vs. text

ALT tags are required by the EPUB3 spec. A file will fail validation if a tag is missing. But a tag can contain no information (no text) and still pass validation. Here's an empty alt tag:

alt="" (Note the quote marks enclosing nothing.)

Alt text, on the other hand, is the actual text within the tag.

alt="image description here"

This provides the necessary information about the image. And that information has to come from the editor, photo editor, or author.

What's the editor's role?

- Provide ALT text in the manuscript so the typesetter can include it in the image tagging.

Ornaments and display typography as image

Decorative images, like ornaments used for narrative pauses, don't need ALT text. The ALT tag itself must be included in the XHTML (that's the developer's job), but it can remain empty with an ARIA role assigned (role="presentation") which indicates it can be passed over and not announced by assistive technology.

An ebook developer will sometimes convert the title page or chapter openers to images to reflect the print design. Whatever text is in that image must be included as ALT text.

Tables and math and scientific notation are also frequently captured as images, so live text is not available for assistive technology. See Chapter 8 for more information on tables.

See Resources in Appendix **D** for links to sites that coach ALT text writing.

Sample ALT text

The text around this image discusses a day hike, but doesn't mention the boardwalk across the meadow.

The caption reads

We hiked through the National Park on Friday.

So, looking at the surrounding text in combination with the caption, we see we may need ALT text to convey the fullness of the image. It could read

Two hikers cross a meadow on a narrow, raised boardwalk.

Video captions

If you're including video—in an instructional manual, for example—you'll need to caption the dialog. It's an accessibility requirement, but it's also helpful for anyone viewing the content in an environment where it's difficult to hear.

There are inexpensive automated captioning services online if a script doesn't exist. A video can also be uploaded to YouTube, which will create a caption that is downloadable as an editable text file. Combining the captioning with the video is a task for the media producer or ebook developer.

KEVIN CALLAHAN

Captions created by automation software often need editorial work. You may have to

- add or adjust punctuation
- correct poor transcriptions
- fix homophones: *you're* to *your*
- delete distracting hmm and umm and uhh

Chapter 11: Ebook QA

What is quality assurance?

Ebooks require quality assurance (QA, or, as most publishing pros call it, proofreading).

Doing ebook QA means that not only are you looking at text but also at

- content completeness
- presentation
- hyperlink viability
- appropriate color use
- metadata

To make QA simpler, settle all editorial issues before EPUB creation. Text and design changes can be handled in the EPUB, but at extra time, cost, and the risk of introducing errors.

Where do I QA?

It's hard to know which devices your readers might use. So, try a "solid read" on one platform and spot checks on another couple.

Keep this in mind: You won't see the final, published rendition of your ebook until you upload it to a retailer and purchase from their site. Retailers process EPUB files to suit their particular needs. Sometimes what you see on a sideloaded Kindle Fire, for example, won't match what you see once a book is purchased from and delivered by Amazon.

QA with devices, apps, and previewers

- Kindle devices (Amazon)
- Kindle Previewer 3 *previews* Enhanced Typesetting, Amazon's refined typography
- Enhanced Typesetting won't be *embedded in* the ebook until upload to Amazon. So if you export a mobi from Kindle Previewer and load it onto a Kindle, it won't look like it does in Previewer. For example, the Bookerly font won't be available. Enhanced typesetting is permanently applied when the file is uploaded to Amazon.
- iPad, iPhone, Macintosh computer for Apple Books
- Nook (Barnes and Noble)
- Kobo devices
- Adobe Digital Editions, a Windows and Mac desktop app, approximates how a book will appear. It's notorious for poor image rendering and malfunctioning links and other oddities.
- Apps for all retailers on Apple and Android devices, Macintosh and Windows computers
- Calibre ebook management

You will find links in Resources.

Get the apps

- Sign up as a customer at a retailer.
- Download their free apps onto your devices and computers.
- Apple Books is already installed on Apple devices.

App caveat

Apps that are installed on phones and computers (as opposed to devices like the Kindle or Nook) are sometimes the last to be upgraded when a device maker changes its software. The Nook app for iPad, for example, went unchanged for years, and performance suffered.

How do I ask for QA changes?

I don't like to lead with a negative examples, but after many years of deciphering change requests, the difficult ones are top of mind.

Do not
- ask to change its to it's on page 23 (Ebooks don't have page numbers; they do have location numbers, but those can change depending on font, font size, device, etc., so your location 23 may be 32 for me.)
- forget to note what device or app you are using
- say something *looks weird*
- request global changes without giving specific examples
- be afraid to ask questions *(Why doesn't this image expand when tapped on screen?)*

Do
- be specific; detail
- chapter number or name
- unique local text snippet
- what devices, apps, or browsers you are using
- give specific new language, not generalities
- be descriptive: *photo looks low-resolution; face seems blurred*
- query the developer for recurring problems instead of detailing each occurrence (It may be fixable via CSS or global find/change.)

Be a sleuth
- do some googling
- ask on Twitter (#eprdctn)
- go to user forums at device websites

The QA checklist

- **Book title and author name:** Find the location on the device/app/browser that mentions the title and author name. Are they correct? Direct the developer to fix the metadata if needed.

- **Tables of Contents:** Check links in both device listing and listings within the flow of the book. Always check again after a text or formatting change.
- **Text:** No need to read the whole book, but certainly spot check; slug paragraphs or major-level heads.
- **Hyperlinks:** Check inside-the-book links like glossary and index links, and those that go to websites outside the book.
- Notice how links appear across devices or previewers. Are they obvious? Do they have underscores? Are they in a different color? If not, ask developer to adjust code.
- **Em (and en) dashes:** They should not appear at the beginning of a line when text reflows. Request that the developer use a nonbreaking space before the dash (* * in markup).
- **Ellipses:** An ellipse should not begin a line when text reflows. There shouldn't be a break inside the ellipse, either. For accessibility, direct the developer to use the XHTML glyph preceded by a nonbreaking space for ellipses rather than typed-out periods with spaces between (as is often done in print). A text-to-speech rendering of the glyph will pause while reading, but those periods will be pronounced, which is probably not the author's intention.
- **Fractions:** Check all fractions. If any break, ask developer to research HTML entities.
- If a fraction does not have an HTML entity, consider expressing as a decimal.
- **Quote marks, apostrophes:** These can become slab quotes, or turn backwards, particularly in OCR scans or PDF-generated text. Check that single and double quotes are facing the right way.
- **Diacritics:** Accents and other glyphs can disappear or turn into gibberish if not inserted properly. Keep track of special characters during print production and check once in ebook. Ask developer to use HTML entities for missing/incorrect characters.
- **Small caps:** Not all devices create small caps well. Keep track of them. Developers often retype text that is meant to be small caps as all caps, then apply code to make them smaller.

- **Drop Caps:** These need different code for different devices. Check each device the book is intended for. If there's a problem, ask the developer to consult requirements.
- Don't use drop caps if poor rendition is a concern; consider a stick-up cap, or bold the first word or line of text.
- **Lists:** Check that numbered and bulleted lists are in order and that hanging indents are correctly indented. The developer can adjust code for misnumbered or poorly aligned lists.
- InDesign can introduce poor markup when exporting to EPUB, so ask developer to double-check that all lists are correct in the XHTML files.
- **Page turning:** Notice how smoothly the turn from one chapter to the next is. If you experience a delay, the new chapter may contain lots of images or hyperlinks or cross-references, all of which can make a file larger and slower. Query the ebook developer, asking if the file can be broken up to improve speed.

Fixed-layout QA

Fixed-layout EPUBS are different from reflowable, so there's a different approach to QA.

Hopefully you'll only be seeing children's picture books or other highly designed, text-light books. Those are the best candidates for this format. Anything with a lot of text in regular text columns should be reflowable.

There are several different tools for creating these ebooks. Some rely on a PDF, others are exported directly from InDesign.

In any case, the fixed-layout EPUB should be a mirror image of the final print document. Fonts and page design should be the same, as should page numbers and artwork.

At the moment, fixed-layout titles are pretty inaccessible; there is often no head hierarchy or available text-to-speech, and images won't have ALT text. Standards and workflows are being discussed now in the governing community, but until then, this list can suffice for QA:

- You can check fixed-layout ebooks in Apple Books, Adobe Digital Editions, and Kindle Previewer. Note that depending on how the file was made, an EPUB may not be viewable on Kindle Previewer. Ask the developer for the appropriate file.
- Check metadata: Is the book title appearing properly in the reading system? Author name?
- Compare each page with the source. Do fonts and line breaks match?
- Are all elements of the print page in the ebook?

Appendix A: XHTML and CSS samples

I'm a great fan of "playing the tape to the end" when setting out on a task. If I know how my work will be implemented down the line, I can be as complete as possible. While most editors won't ever need—or want—to see inside an EPUB, it can be helpful to have an idea what the components are and how they work together.

This is not an exhaustive survey of XHTML and CSS implementation best practices. For that, there are lots of websites and books and training videos online.

Read on for an overview. If writing good markup strikes your fancy (and I'm convinced my background as a copyeditor helped me parse XHTML when I first encountered it), then by all means, learn it!

The XHTML file

Books are typically divided into chapters, and so are EPUBs. One instance where this may not apply is if a chapter is very long or has a lot of images and becomes too large for some e-reading devices to navigate smoothly. The developer may break a chapter into multiple pieces.

XHTML Markup

An XHTML file contains markup: basically, code written to be read by a machine, an ebook-reading system in this case. In XHTML, every block of text is marked with a beginning and an ending tag. So the beginning of a chapter may look like this:

```
<h1>Chapter One: The Name of the Chapter</h1>
<h2>A Subhead</h2>
<p>The first paragraph under the subhead begins here.</p>
<p>This is the second paragraph.</p>
<blockquote>An extract is enclosed in the blockquote tag</blockquote>
```

`<h1>` is the chapter number and title.
`<h2>` is the first subhead in the chapter.
`<p>` is the paragraph tag.
`<blockquote>` is an extract.
Every subsequent chapter would have a similar structure. Someone using assistive technology would be able to move through the book by hearing all the `<h1>` entries.

Classes

Classes introduce a variation to an existing paragraph tag. The markup in this instance:

```
<p class="left">Read about the best book ever.</p>
```

creates a flush-left paragraph. The definition of `class="left"` is found in the CSS:

```
.left {
text-align: left;
}
```

Semantic markup

Say you want to emphasize the word *ever*:

```
<p class="left">Read about the best book
<em>ever</em>.</p>
```

And if you want to add a trademark symbol to the word *ever*:

```
<p class="left">Read about the best book
<em>ever&#174;</em>.</p>
```

where ® indicates a trademark symbol ™ expressed as XHTML code. You want the code so a screen reader knows to pronounce *trademark*. If you typed out *TM*, that's what the screen reader would say out loud.

You can see why it's preferable to finish editing and proofreading before creating the EPUB. There are enough moving parts—all the opening and closing tags—that it wouldn't be difficult to mistakenly delete or damage an opening or closing tag.

Appendix B: Accessibility checklist

I've tried to fold accessibility ideas into the general conversation because they should always be kept in mind while editing. But here's a list of concepts that you can refer to for quick refreshers.

- Head levels: are they consecutive? Are any missed (i.e., h1 followed by h3)?
- Italics and bold: are the different varieties of each differentiated?
- Color: are combinations and contrast legible? Do they meet the WCAG standards?
- Hyperlinks: are they underscored and rendered in a different color than the text?
- Link rot: are external hyperlinks viable? Is there a solution for keeping URLs current?
- Languages: is the base language specified? Are there exceptions, and are they marked?
- Navigation: are landmarks noted? ARIA roles?
- Images: is ALT text included for images that are not described in text and/or caption?
- Video: are they captioned? Do the captions need editing?
- Tables: are complex tables rewritten so they can be rendered as easily understood as live text?

Appendix C: Transmittal to the ebook developer

Here's a list of items to mention to the book designer, typesetter, and/or ebook developer. Some might be obvious to anyone looking at a manuscript or typeset document, but it can't hurt to point them out so they aren't missed in the ebook-making process. Keep in mind that the typesetter and ebook developer might well be two different people, so don't assume communicating something to the typesetter will find its way to the ebook maker.

Think of the transmittal as similar to creating a design survey, where you pull out all the distinct elements of a manuscript for the designer to address. Don't forget:

- diacritics
- foreign languages
- fractions
- math/science notation
- small caps
- hierarchy of heads: how to treat part/chapter number and part/chapter titles
- ebook-only elements
- color images vs. b/w images
- elements to exclude from print but retain for ebook use

- landmarks
- list all major book elements and their corresponding Landmark titles
- ARIA role and epub:type designations
- devise a system for marking this in the manuscript or separately

Appendix D: Resources

Accessibility

Accessibility definitions
   3playmedia.com/2018/08/23/accessibility-terms-defined/h

An Accessibilty Statement from Macmillan Learning
   https://www.macmillanlearning.com/college/us/our-story/accessibility

"Born Accessible"?
   booknetcanada.ca/blog/2019/6/20/producing-born-accessible-books

The Daisy Consortium
   https://daisy.org

EPUB Accessibility Definitions
   kb.daisy.org/publishing/docs/conformance/epub.html

Free Webinars on multiple accessibility topics, including multiple sections on creating descriptive text
   https://daisy.org/webinar-series/

General ideas about accessibilty
   ericwbailey.design/writing/truths-about-digital-accessibility.html

Some tips and ideas
   deque.com/blog/accessibility-strategies-for-your-content-team/

What ebook features are popular?
   booknetcanada.ca/blog/2019/6/20/producing-born-accessible-books

Accessibility validation

While EPUB validation is essential to upload a file to a retailer or distributor, accessibility validation is not. However, accessibility is such an important feature of ebooks that it's a valuable step to take.

ACE, by Daisy

ACE is a desktop application that will check an EPUB's accessibility features: head hierarchy, metadata, image ALT tags.

youtube.com/watch?v=qEs0r2hwuLY

Benetech Global Certified Accessible Certification

Benetech is dedicated to making all web-based content accessible and available to as many readers worldwide as possible. Their Global Certified Accessible program ensures that a publisher's ebooks meet current accessibility standards. Benetech will work with a publisher to establish in-house workflows and standards.

bornaccessible.benetech.org/

ALT text

Write ALT tags yourself

Learn how to write ALT text:
Diagramcenter.org/poet.html
Learn about when long descriptions are needed.
diagramcenter.org/best-practices-for-authoring-extended-descriptions.html

Hire someone to write your ALT tags

These vendors provide ALT text writing services.
TextBox Digital: textboxdigital.com/
247 Accessible Documents: 247accessibledocuments.com/2020/10/21/top-5-resources-to-help-you-write-better-image-descriptions/

ARIA roles

An up-to-date list is here.
w3.org/TR/dpub-aria/ - roles

Book Industry Study Group
The BISG has several guides about digital publishing.
   bisg.org/general/custom.asp?page=Guides

Calibre ebook management
   calibre-ebook.com/

Chicago Manual of Style
While the CMOS doesn't dedicate a unique section to ebooks, references and best practices are scattered throughout CMOS17.
   chicagomanualofstyle.org

Color contrast checker
This is a designer's task, but it's helpful for an editor to be familiar with WCAG 2.1 requirements. Numerous sites provide simple tools to check text color against background color. Here's one:
   webaim.org/resources/contrastchecker/

Ebook accessibility newsletter
Monthly newsletter about international accessibility news and book publishing in general.
   inclusivepublishing.org/

Ebook knowledge base
Information on ARIA roles, semantics, and other specs:
   kb.daisy.org/publishing/

Ebook-building platforms
A growing number of tools are appearing that will produce print-on-demand-ready PDFs and ebooks from a text document. Their developers are smart and dedicated to accessible EPUB3 standards.

Bookalope
   bookalope.net/

Hederis
   hederis.com/

IngramSpark Book Builder
   ingramspark.com/

Vellum
   https://vellum.pub

If you're looking to get your hands on some XHTML and CSS, these two apps get you under the ebook hood.

Jutoh
   http://www.jutoh.com

Sigil
   https://sigil-ebook.com

EPUBsecrets
Insider's look at how ebooks are made and the issues that arise.
   epubsecrets.com

EPUB validation
An EPUB must be valid to be uploaded to any retailer or distributor. The most commonly used tools are

Pagina epub-Checker
This is a free application that many ebook developers use on a daily basis. It is usually the most up-to-date with standards and best practices.
   pagina.gmbh/produkte/epub-checker/

Ebook FlightDeck
FlightDeck provides an excellent, free Handbook. The validation service is subscription based.
   https://www.ebookflightdeck.com

Hyperlink viability (link rot)
Has link rot set in? Here are some ideas for dealing with it.
   epubsecrets.com/when-good-links-go-bad-link-rot-in-ebooks.php

This service will create an archive of URLs so they don't go out-of-date.
   archive.org/web/

Index
LiveIndex is an InDesign script that will link the page reference to the spot in the text that corresponds to the top of the print page. It also creates linked indexes for PDF presentation.
   id-extras.com/products/liveindex/

Kindle Previewer
This is free, and mostly accurate to what you'll see in a book once uploaded to Amazon.
   amazon.com/gp/feature.html?ie=UTF8&docId=1003018611

Landmarks
Description of use and list of landmarks.
   idpf.github.io/epub-vocabs/structure/

Learn web/ebook technology
One good resource is the W3C Schools.
   w3cschools.com.

Math, science notation
This is a difficult area to discuss fully in a short booklet, so here are fuller explanations of MathML.
   w3.org/Math/
   en.wikipedia.org/wiki/MathML

Semantics
Why and how we use semantics.
   lifewire.com/why-use-semantic-html-3468271

W3C Publishing Working Group
These are the folks who write the specs.
   w3.org/publishing

Glossary

A11Y: shorthand for accessibility. The acronym is formed by the first and last letters, with the number 11 representing the number of letters between the two.

ACE, by Daisy: Free application that validates an EPUB's accessibility features. It checks metadata, head hierarchy, presence of ALT tags, and more. ACE is an app that sits on your computer's desktop. To operate, drag an EPUB into the open application and wait a minute or two. You'll see a screen with any issues found. You can download a report to share.

ARIA role: designation of a book element, such as part, chapter, index. ARIA roles were developed for use in building accessible websites, but there are analogous ebook–specific roles. The acronym stands in for Accessible Rich Internet Application.

Assistive Technology (AT): Software that aids readers through text-to-speech, navigation, and other tools.

CSS: Cascading Style Sheets. Contains presentation attributes for XHTML documents, such as font, color, position, and alignment.

Color Contrast: In ebook terms, how readable text is against a color or tint. Very light text set against very light gray could be difficult to read on some ebook devices, and impossible to read for some users with vision problems. The accessibilty of color contrast can be checked using an accessibilty validator that uses WCAG 2.1 standards (see below).

Daisy Consortium: An international organization that works with publishers, tech developers, and device and app creators to ensure accessible reading for people with disabilities. Daisy created the accessibilty validation tool, ACE.

embed: include items in an EPUB. Typically refers to fonts, which can be included in ebooks as long as they are specifically licensed for ebook use.

EPUB: The source file of an ebook. Based on web technology, it shares with web sites XHTML documents that contain the content, CSS documents that include presentation instructions, images, fonts, audio and video files, and javascript for animation. An EPUB must include ebook–specific files that complete a self-contained package, such as the content.opf and the mimetype. While a collection of XHTML and CSS files can generally be viewed on any web browser, EPUBs can only be viewed on dedicated EPUB–reading software and devices such as a Kindle, a Nook, or Apple Books.

EPUB-Checker: a free, downloadable EPUB validation application. It is kept current with the EPUB specification. You can either upload a file (maximum size: 10MB), or download a desktop app. It makes sure all required pieces of the EPUB are present, and checks for valid XHTML files. Developers can provide a report to clients proving a file is valid.

epub:type: a system to mark book elements, such as part, chapter, index. Designations are parallel to ARIA roles. While defunct in practical use within assistive technology, epub:type is useful because developers can assign them within InDesign and, after export, change to the equivalent ARIA role.

Export: use layout or text editing software to create an EPUB. InDesign is the most common layout software used.

FlightDeck: online, subscription-based EPUB validation tool. The site includes a free handbook with a lot of helpful information. You can upload an EPUB to validate, add metadata, check against various retailer requirements. The site also provides a list of all images and hyperlinks in an EPUB.

Landmarks: accessibility navigation feature that lists major sections of a book, such as frontmatter and backmatter, and more granular sections, such as Introduction, Chapter, and Index.

Markup: the XHTML file that contains content that is tagged (h1, h2, p, etc.) semantically, according to its function in the text.

Math ML: a web–based markup tool for complex mathematic equations.

MOBI: Amazon version of an EPUB, typically created once the EPUB is complete. Until 2020 it was common for publishers to upload a MOBI to Kindle Direct Publishing, but Amazon currently requests that an EPUB be uploaded.

multimedia: usually refers to audio and video in an ebook. Reading device support is not universal, so caution is needed when including.

Navigation: provides hyperlinked routes around an ebook. Can be via Landmarks to bring you to the backmatter, or the device table of contents to hop to different chapters, or the Page List, which brings you to the spot that matches the print edition's pagination.

Page List: a hyperlinked list that brings you to the spot in the ebook that matches the top of a particular print page. Helps coordinate between print and ebook readers in a setting such as a classroom or a book group. Include which print edition the Page List is based on within the metadata.

Print-on-Demand (POD): the ability for a service, such as Kindle Direct Publishing or Ingram Spark, to print books based on sales needs. A PDF of a print layout can be uploaded to the service, or in some cases a text file.

Quality Control: checking an ebook for completeness of content, functionality of hyperlinks (cross-references, footnotes, index, as well as links to websites outside the ebook), functionality of tables of contents. It's recommended to look through an ebook using different devices at different font settings, using all the variables inherent in an e-reader. Frequently referred to as Quality Assurance.

Semantic: correct formatting of an XHTML document, using universal tags to mark copy as headings, text, figures, figure captions, tables, and more. Semantic markup is essential for cross-platform functionality of an EPUB file, for devices and apps that exist today and those that will be invented in the future.

Sideload: load an ebook onto a device, usually using a USB connection between, for example, a Kindle and a PC. Often used for doing quality control on an ebook before distributing to retailers.

Validation: ensures that an ebook contains all necessary parts and that it is well made, without errors in markup or file structure. Retailers will refuse ebooks that don't pass validation. epub-Checker and FlightDeck are two validation tools. Some ebook–making applications, like SIGIL, include validation features as well.

WCAG (Web Content Accessibilty Guidelines): Developed and maintained by the W3C, WCAG provides guidance on ensuring that content (text, images, audio, video) is accessible for all users. It includes all accessibilty issues: text alternatives, video captioning, and use of color and color contrast. The guidelines pertain to both web and ebook development. As of this writing, Version 2.1 is in force. There are three levels of WCAG compliance: A, AA, AAA. An example: Level A (the most basic, which all publications should meet) calls for the use of alternative text for images and for correct heading hierarchy. Level AA includes minimum color contrast requirements. Level AAA provides guidance on total keyboard navigation for a document.

World Wide Web Consortium (W3C): develops web and ebook specifications. This means file structure, best practices, accessibilty features.

XHTML: a file that contains text and links to image, audio, and video files that is used by web pages and ebooks. It has strict rules for semantic structure. An ebook often contains an XHTML file for each chapter. This content file is linked to the CSS, which is the style sheet that governs presentation.

About the Author

Kevin Callahan owns BNGO Books, a New York City–based ebook and print design company. BNGO adapts ebooks for publishers as diverse as Scholastic, the Associated Press, the American College of Emergency Physicians, and the Mayo Clinic. Kevin writes frequently about digital publishing and presents webinars and in-person training on ebooks and accessibility. He is a former editor-in-chief of epubsecrets.com and has served on the faculty of Pace University's Masters in Publishing program. His course on creating fixed-layout ebooks for the Kindle is on LinkedIn Learning/Lynda.com. You can follow him @BNGOBooks and see his work at **bngobooks.com**.

About the
Editorial Freelancers Association (EFA)

Celebrating 50 Years!
Dedicated to the Education and Growth
of Editorial Freelancers

The EFA is a national not-for-profit — 501(c)6 — organization, headquartered in New York City, run by member volunteers, all of whom are also freelancers. The EFA's members, experienced in a wide range of professional skills, live and work all across the United States and in other countries.

A pioneer in organizing freelancers into a network for mutual support and advancement, the EFA is now recognized throughout the publishing industry as the source for professional editorial assistance.

We welcome people of every race, color, culture, religion or no religion, gender identity, gender expression, age, national or ethnic origin, ancestry, citizenship, education, ability, health, neurotype, marital/parental status, socio-economic background, sexual orientation, and/or military status. We are nothing without our members, and encourage everyone to volunteer and to participate in our community.

The EFA sells a variety of specialized booklets, not unlike this one, on topics of interest to editorial freelancers at the-efa.org.

The EFA hosts online, asynchronous courses, real-time webinars, and on-demand recorded webinars designed especially for freelance editors, writers, and other editorial specialists around the world. You can learn more about our Education Program at the-efa.org.

To learn about these and other EFA offerings, visit the-efa.org and join us on social media:

Twitter: @EFAFreelancers
Instagram: @efa_editors
Facebook: editorialfreelancersassociation
LinkedIn: editorial-freelancers

www.ingramcontent.com/pod-product-compliance
Lightning Source LLC
Chambersburg PA
CBHW071537080526
44588CB00011B/1698